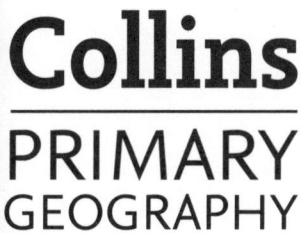

Issues
Workbook 6

Planet Earth

Restless Earth
Earthquakes and volcanoes — 2
Creating landscapes — 4
Rocks and soils in the UK — 6

Water

Drinking water
Water, water, everywhere — 8
Water supplies — 10
Conserving water — 12

Weather

Climate change
Global warming — 14
Unusual weather — 16
Responding to climate change — 18

Settlements

Planning issues
Reasons for development — 20
Old sites, new uses — 22
Planning game — 24

Work and Travel

Transport
Travelling further, travelling faster — 26
Transport problems — 28
Hidden costs — 30

Environment

Conservation
Threatened wildlife — 32
Antarctica — 34
Conservation projects — 36

Places

England — 38
Europe — 44
South America — 50
Asia — 56

Fiona Macgregor

Unit 1 Restless Earth

Lesson 1: Earthquakes and volcanoes

1 a) Draw lines to match the words to the descriptions.

Earth's crust

mantle

core

very hot melted rocks that can flow like a sticky liquid

an extremely hot ball of iron and nickel

the surface of the Earth, made of solid rock

b) Label the diagram using the words from part **a)**.

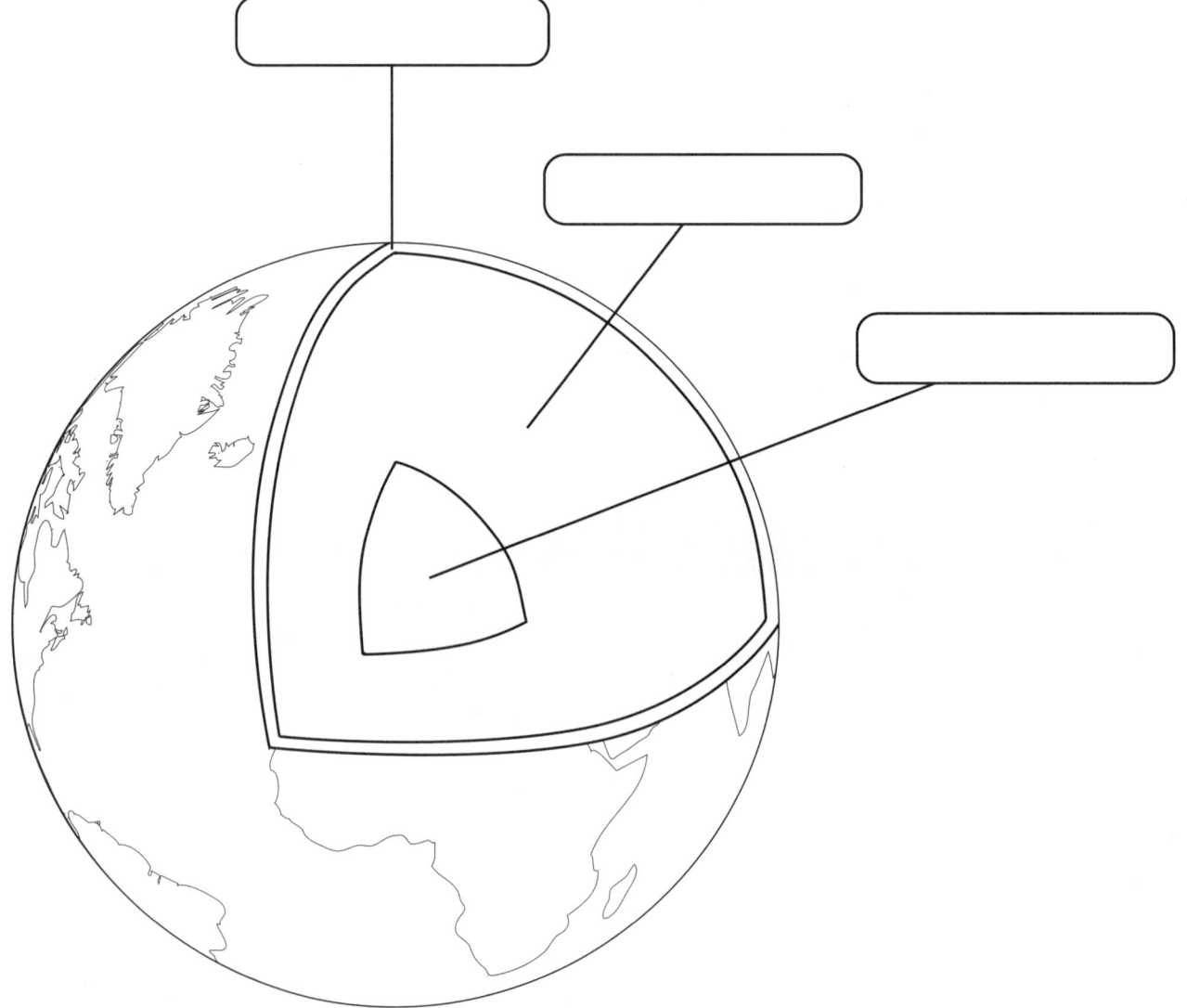

Unit 1 Restless Earth

② Look at the diagram of the volcano below.

   a) Colour in the key.

   b) Now colour the diagram to match the key.

   Hint: Magma is molten (very hot, melted) rock deep below the Earth's surface. When magma erupts from a volcano, it is called **lava**.

   Key

magma	red
lava	orange
ash cloud	grey

   c) Label the diagram. Use the words in the key.

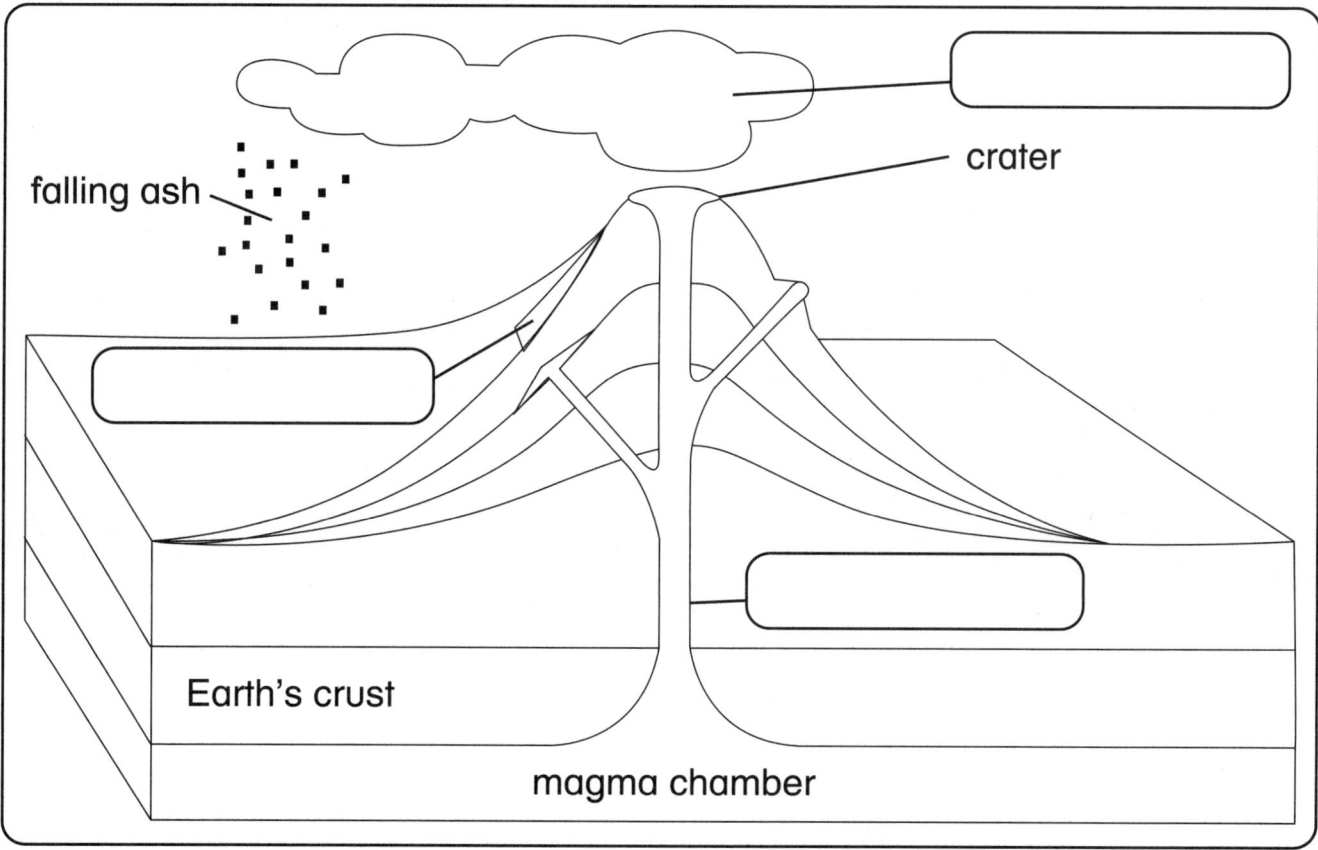

③ Arrange these sentences in order to explain how a volcano erupts. Write the letters in the correct order in the boxes below.

   A When the volcano erupts, it forms a crater at the surface.

   B The lava flows out of the crater and down the outside of the volcano.

   C Molten magma pushes up from the magma chamber.

   D The magma moves up towards the surface.

   E Clouds of ash also escape and ash falls when the volcano erupts.

C					

→ Supports Pupil Book Investigation, page 3

Unit 1 Restless Earth

Lesson 2: Creating landscapes

1 Look at these photographs. For each photograph, write a sentence saying how the shape of the rocks formed.

Use the words from the boxes to help you. You can use the words more than once.

(frost) (headlands) (rain) (winds) (ice)
(waves) (water) (dissolves) (rocks)

Cliffs form when _____

Rocks break apart when _____

Glaciers _____

Caves form when _____

4

Unit 1   **Restless Earth**

② a) Choose a type of rock formation, like the ones shown in Activity 1.
- Draw a diagram to show how the rock was formed.
- Use arrows and labels.
- Give your rock formation diagram a title.

b) Write a sentence saying how the rocks were formed.

_____

_____

Unit 1 — Restless Earth

Lesson 3: Rocks and soils in the UK

1 a) Label the rocks used to build the house and road. Use the words from the boxes.

brick   clay   flint   granite

b) Draw a picture of a house you know. Can you label any rocks used to build it?

Unit 1   **Restless Earth**

❷ a) Collect two different rocks from your local area.

b) Examine them carefully and then complete this table.

	Rock specimen 1	**Rock specimen 2**
Where I found it		
What colour it is		
What it feels like (for example: crumbly, smooth, rough, sharp)		
What I think this type of rock could be used for (for example: buildings, pavements, roofs) and why I think so		
A sketch of the rock		

Unit 2 Drinking water

Lesson 1: Water, water everywhere

1 Look at the diagram showing how water is sourced and stored in the UK. Match the numbers on the diagram to the descriptions in the boxes.

- water stored in a reservoir ☐
- rain falls in the mountains ☐
- waterworks: water taken out of the river ☐
- river begins as a stream ☐
- pumping station: water pumped out of the ground ☐

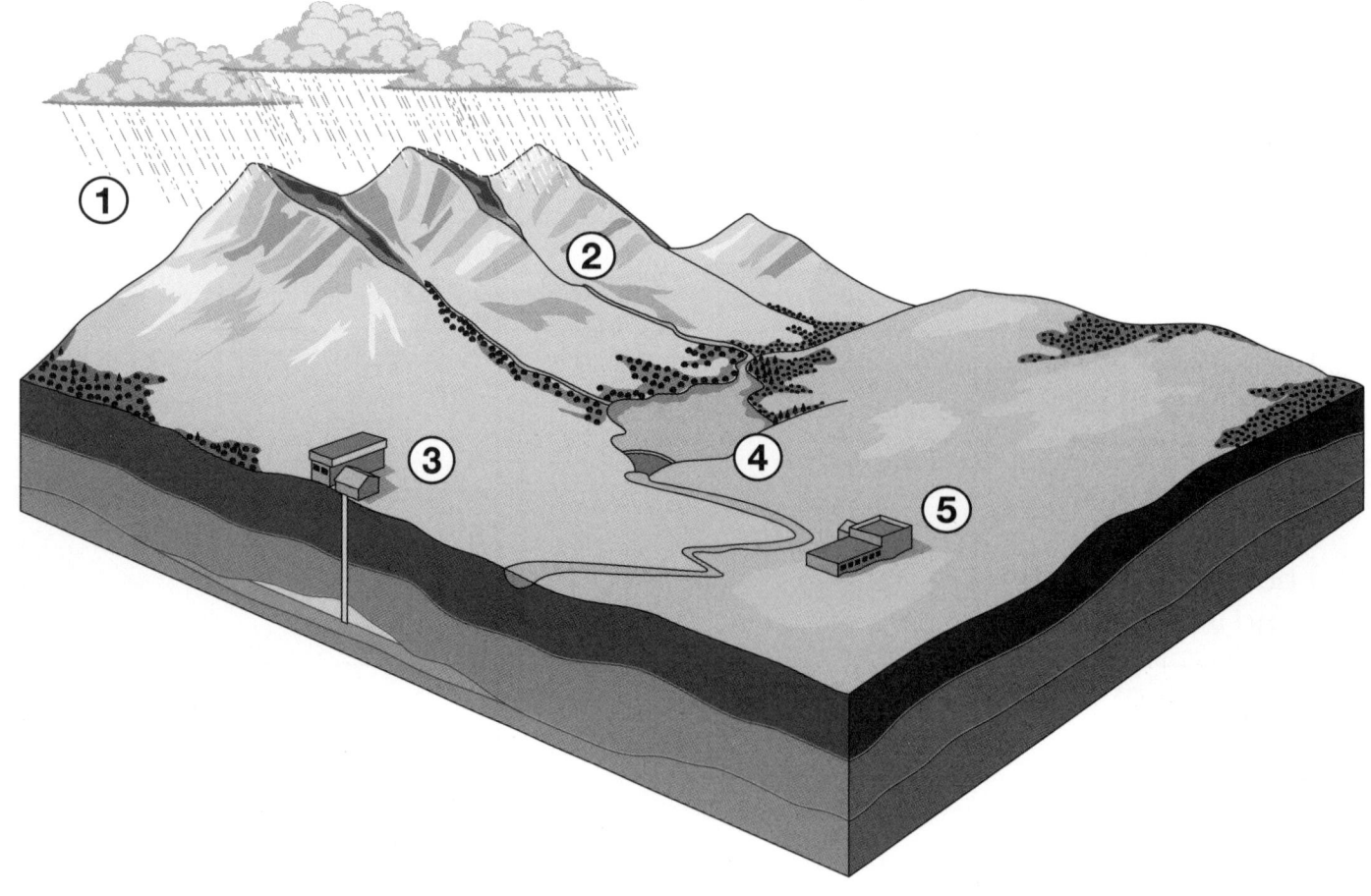

Unit 2   Drinking water

❷ a) Find out about water sources in your local area. Are there any of the following sources? Tick ✓ the ones you have.

Write their names, if you know them.

☐   rivers _____

☐   lakes _____

☐   reservoirs _____

☐   seas _____

☐   streams _____

Hint: Use a map of your local area, ask an adult or do some research.

b) Find or draw pictures of four local water sources. Add them to the boxes.

c) Add captions to your pictures to describe what they show.

Unit 2 Drinking water

Lesson 2: Water supplies

1 Read this extract about the invention of the Hippo Roller.

> The Hippo Roller was invented in 1991 by Pettie Petzer and Johan Jonker, who lived in South Africa.
>
> This clever design means that water is put inside a 'wheel' that can be pushed or pulled along the ground, rather than being carried on someone's head. You can carry 90 kg of water in this way. People who often spend up to five hours a day getting water, can easily pull or push the Hippo Roller over uneven ground.
>
>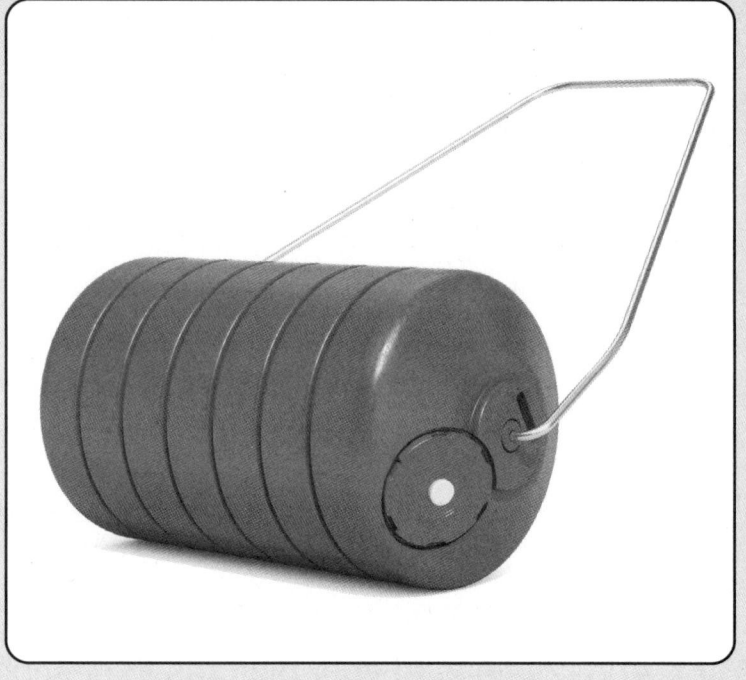
>
> Hippo Rollers are used in over 50 countries in Africa. They are also used in Romania, Papua New Guinea and the Marshall Islands.

2 Answer these questions.

a) Why do you think Pettie and Johan invented the Hippo Roller?

_____

_____

b) Explain how the Hippo Roller works.

_____

_____

c) Find some of the places where the Hippo Roller is used on a map of the world. What might these places have in common? Write your ideas.

_____

_____

_____

_____

Unit 2 Drinking water

3 Read the information in the box.

> **Imagine:** You have to collect water for your family, every day, from a well that is two kilometres away.
>
> **The problem:** Water is heavy, and it is tiring to carry buckets that far every day.
>
> **The solution:** Invent a water carrier.

Now draw your design for a water carrier.

- Label each part.
- Make notes explaining how your invention works.
- Give your invention a title.

Unit 2 Drinking water

Lesson 3: Conserving water

1 Use the information on page 12 of your Pupil Book to complete the fact sheet on the River Indus.

Fact sheet: The River Indus
River source: _____
Length of river: _____
Landscape it flows through: _____
Sea that the river flows into: _____
The past: How the river helped Harappa: _____
Today: The river supports: • _____   • _____   • (green energy) _____   Problems the Indus is facing: _____

Unit 2   **Drinking water**

❷ Complete the poster about how you can conserve water at home and at school. Include six points and six drawings.

Give the poster a title.

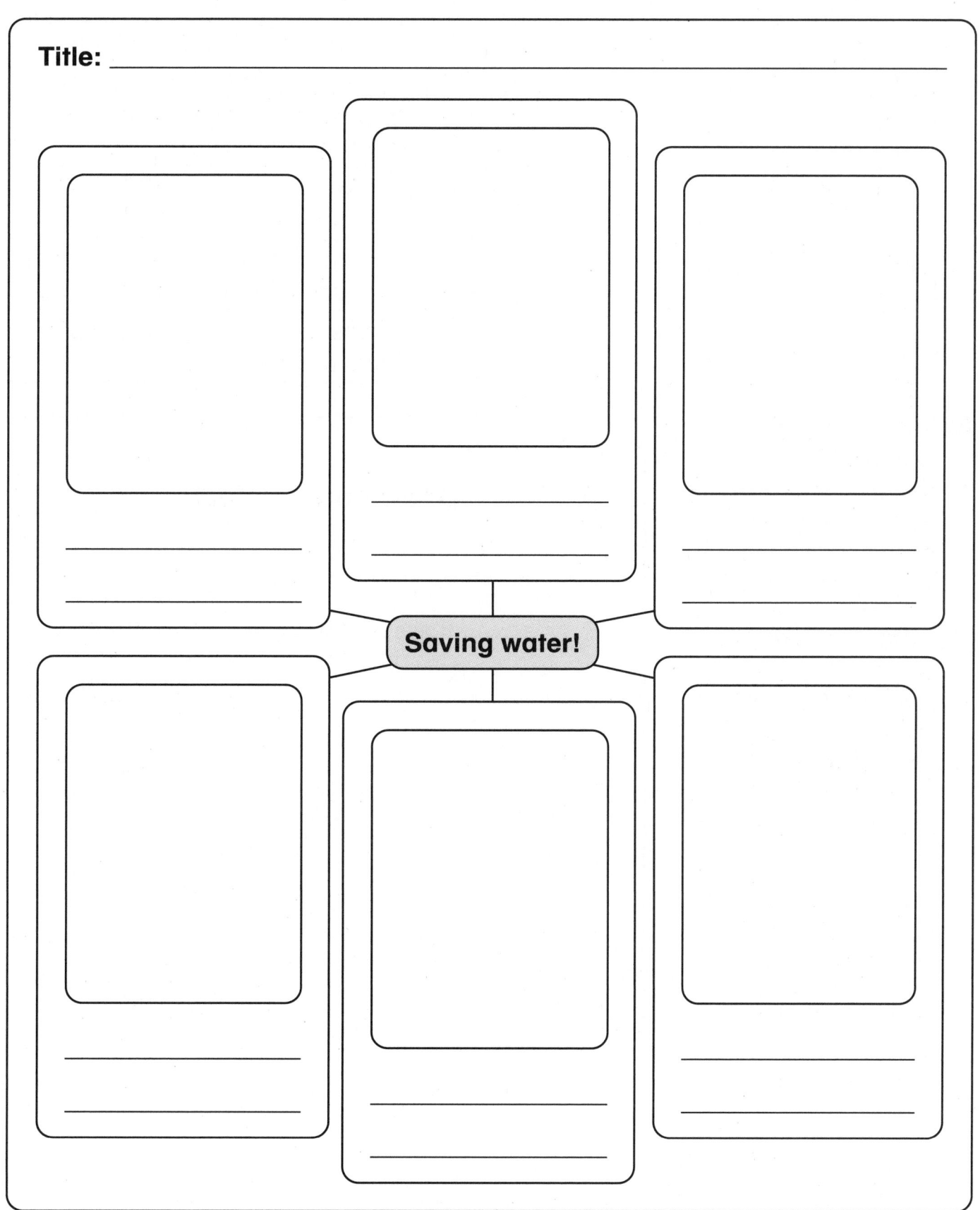

Unit 3 — Climate change

Lesson 1: Global warming

❶ Draw lines to match these words and phrases to their meanings.

climate change	a gas that is given off when fossil fuels are burned
global warming	when heat is trapped near the Earth's surface by greenhouse gases
carbon dioxide	long-term changes in the climate all over the world
greenhouse effect	fuels, like coal, gas or oil, that were formed in the past, and of which there is a limited supply
fossil fuels	the average increase in the temperature of the Earth's atmosphere

❷ Look at this diagram. Complete the sentences to describe what the diagram shows, using the words from the boxes.

atmosphere **warms** **space** **greenhouse** **sun**

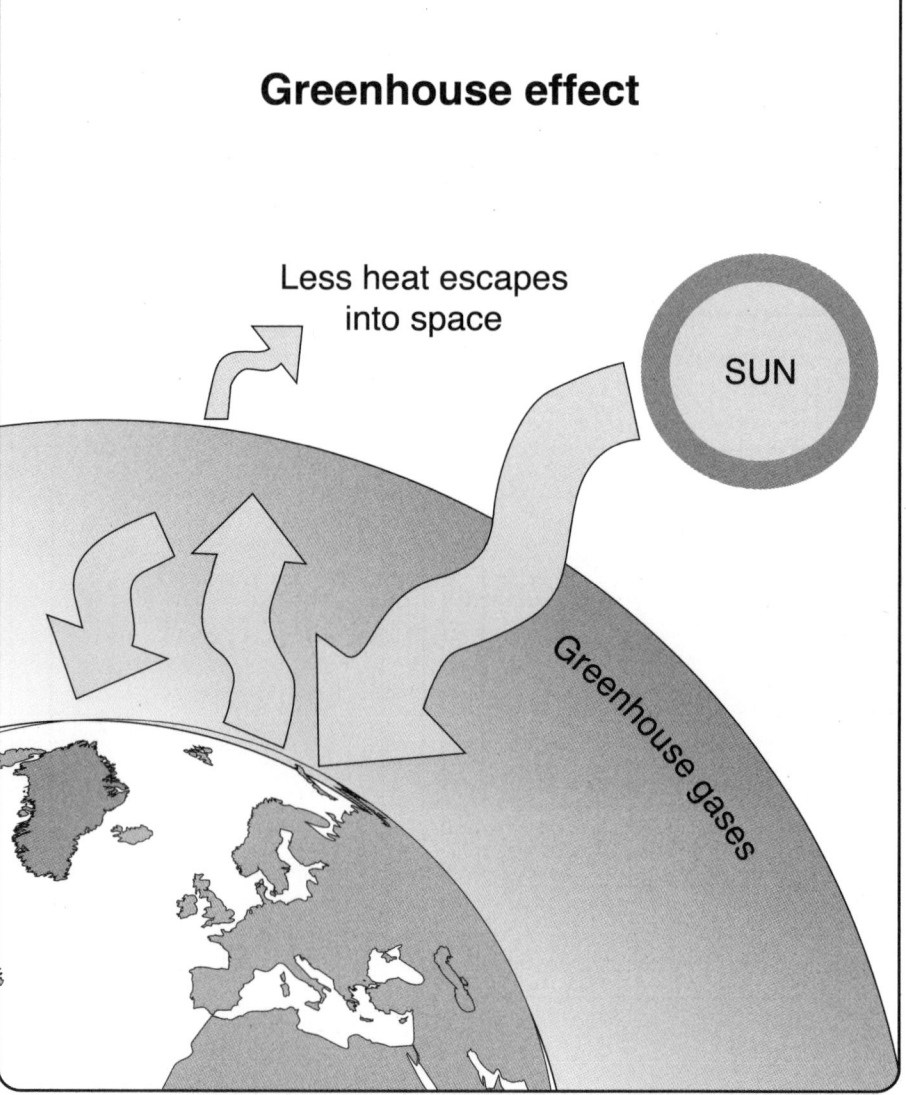

Greenhouse effect

Less heat escapes into space

SUN

Greenhouse gases

_____ gases build up in the _____ and trap heat from the _____. Less heat can escape into _____ so the Earth's atmosphere _____.

Unit 3 Climate change

3 Look at the heat map and information above it on page 14 of your Pupil Book.

   a) Which regions of the world are warming the fastest according to the heat map? Circle the answer.

   deserts   polar regions   forests   seas

   b) People produce greenhouse gases, and they live on the land, rather than at sea. Is there a link to the places that are warming faster? Write a sentence.

   _____

   _____

   _____

   → Supports Pupil Book Mapwork and Discussion, page 15

4 Look at this photograph. What could happen if the Greenland ice sheet collapses? Write your ideas.

- Think about sea levels.
- How could this affect islands and low-lying land?

_____

_____

_____

Unit 3 Climate change

Lesson 2: Unusual weather

1 a) Label the pictures using the words from the boxes.

> drought   forest fire   storm

b) Research recent examples of flooding, drought, forest fires and storms from around the world. Where and when did they happen?

_____
_____
_____
_____

2 Look at a map of the world. List ten small island nations and the oceans or seas around them.

1 _____   6 _____
2 _____   7 _____
3 _____   8 _____
4 _____   9 _____
5 _____   10 _____

→ Supports Pupil Book Mapwork, page 16

Unit 3 | **Climate change**

❸ Research one of the animals on page 16 of your Pupil Book.

Make notes to prepare a presentation about the animal. Include this information:

- where the animal is found
- why it is at risk
- ideas of ways to save the animal.

Hint: Think about how you will order your information. You could present your notes in a diagram to make this clear.

→ Supports Pupil Book Investigation, page 16

Unit 3 Climate change

Lesson 3: Responding to climate change

1 Select three items from this picture. Write how each one contributes to climate change.

_____

_____

_____

_____

_____

2 What is happening in your local area to reduce carbon emissions, plastic usage, or waste?

a) Do some research and make notes below.

_____

_____

_____

_____

_____

_____

_____

_____

_____

b) Use your research from part **a)** to prepare a presentation for your class. You can do this as a talk or as a written presentation. Find photographs or pamphlets to support your presentation.

Unit 3 | **Climate change**

❸ Using the information on pages 18 and 19 of your Pupil Book, write five interesting questions about reducing emissions and reaching net zero in your community. For example: *What can we do to reduce food waste?*

a) Write your questions below, leaving space for the answers.

Question 1: _____
_____
Answer: _____
_____

Question 2: _____
_____
Answer: _____
_____

Question 3: _____
_____
Answer: _____
_____

Question 4: _____
_____
Answer: _____
_____

Question 5: _____
_____
Answer: _____
_____

b) Now swap with a friend and answer each other's questions.

Unit 4 — Planning issues

Lesson 1: Reasons for development

❶ Find an old map of your local area and look at how land use has changed.
Circle the answer you think is true.

Farming:	increased	decreased	stayed the same	none
Housing:	increased	decreased	stayed the same	none
Leisure:	increased	decreased	stayed the same	none
Transport:	increased	decreased	stayed the same	none
Environment:	increased	decreased	stayed the same	none
Industry:	increased	decreased	stayed the same	none

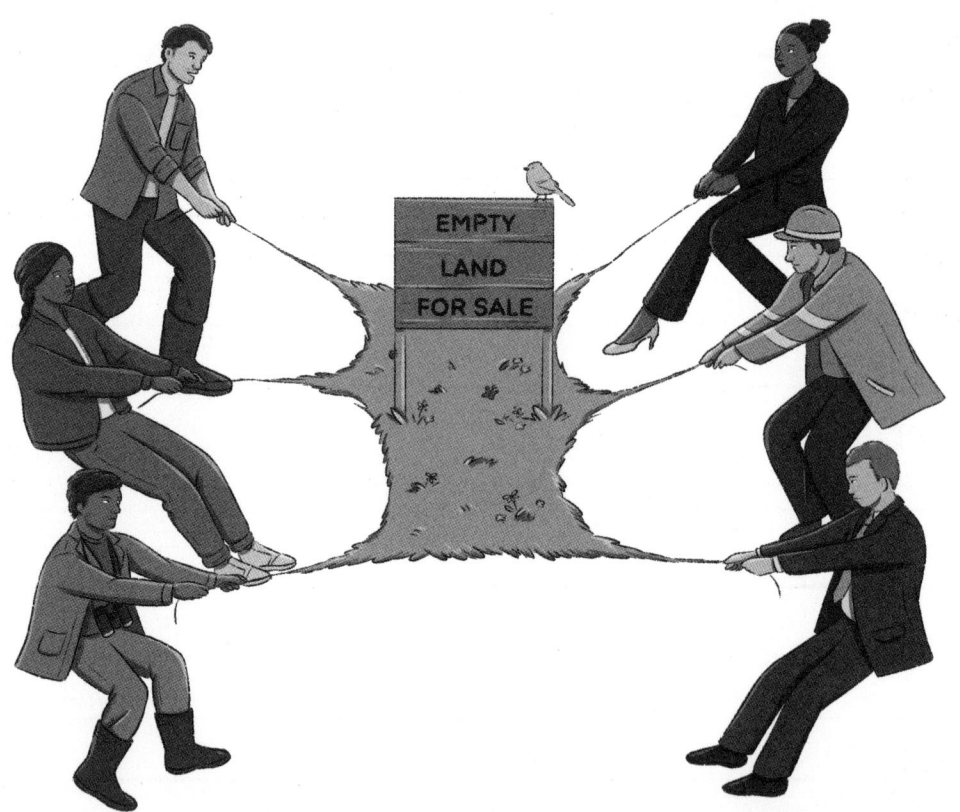

→ Supports Pupil Book Investigation, page 20

❷ Find out if there are any new plans for developing land near you.
Describe the plans.

_____

_____

_____

Unit 4 Planning issues

3 Look at this map of Malta.

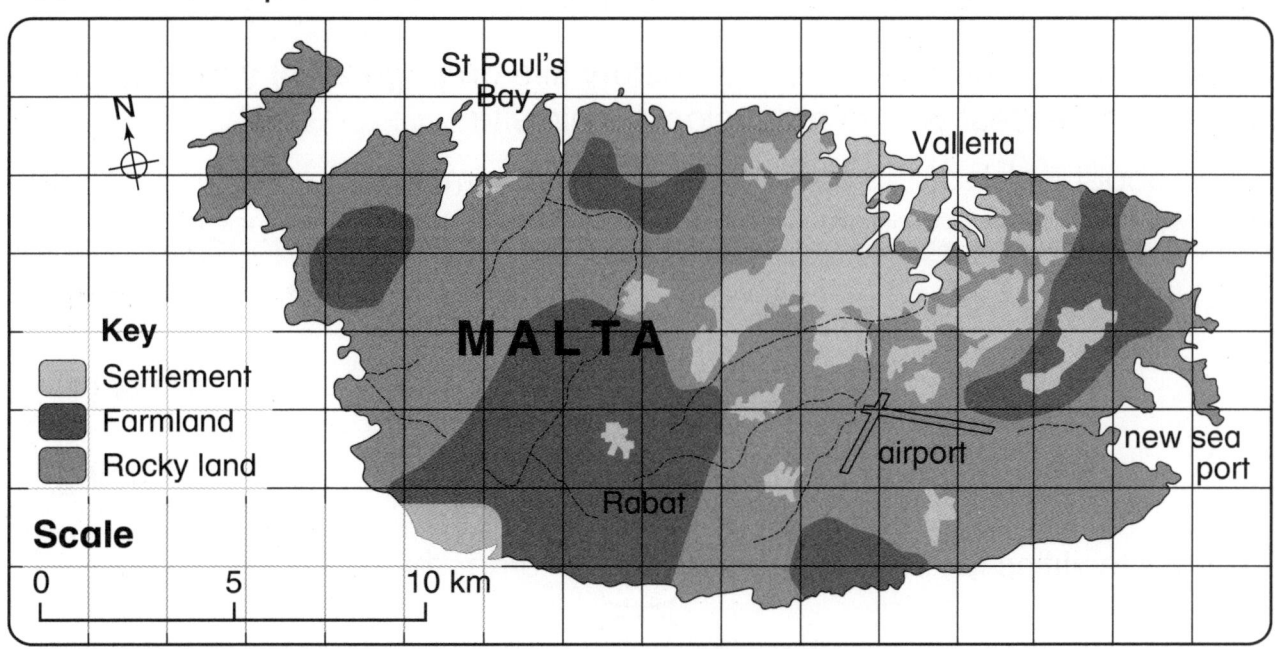

a) Approximately how many squares on the map contain at least some:

   i   settlement _____

   ii farmland _____

   iii rocky land _____?

b) If you were a farmer, where would you put your farm?

   i   Mark it on the map. Use a symbol to represent a farm.

   ii Explain why you put your farm there.

   _____

   _____

   _____

c) If you were going to open a restaurant, where would you put it?

   i   Mark it on the map. Use a symbol to represent a restaurant.

   ii Explain why you put your restaurant there.

   _____

   _____

   _____

Unit 4 Planning issues

Lesson 2: Old sites, new uses

1 Look at the plan of the Cowley car factory site on page 23 of your Pupil Book. Decide on the best way to redevelop the old site.

Think about which of the options below is most important and give each one a score. 1 = most important, 2 = quite important, 3 = least important

local jobs – the factory, offices, etc. ☐   hotels for visitors ☐

shops ☐   transport ☐

the environment ☐   farming ☐

leisure facilities for the community ☐   other things: _____ ☐

housing ☐

2 Now plan the site. You can add buildings, trees and roads. Create a key for the different types of land use.

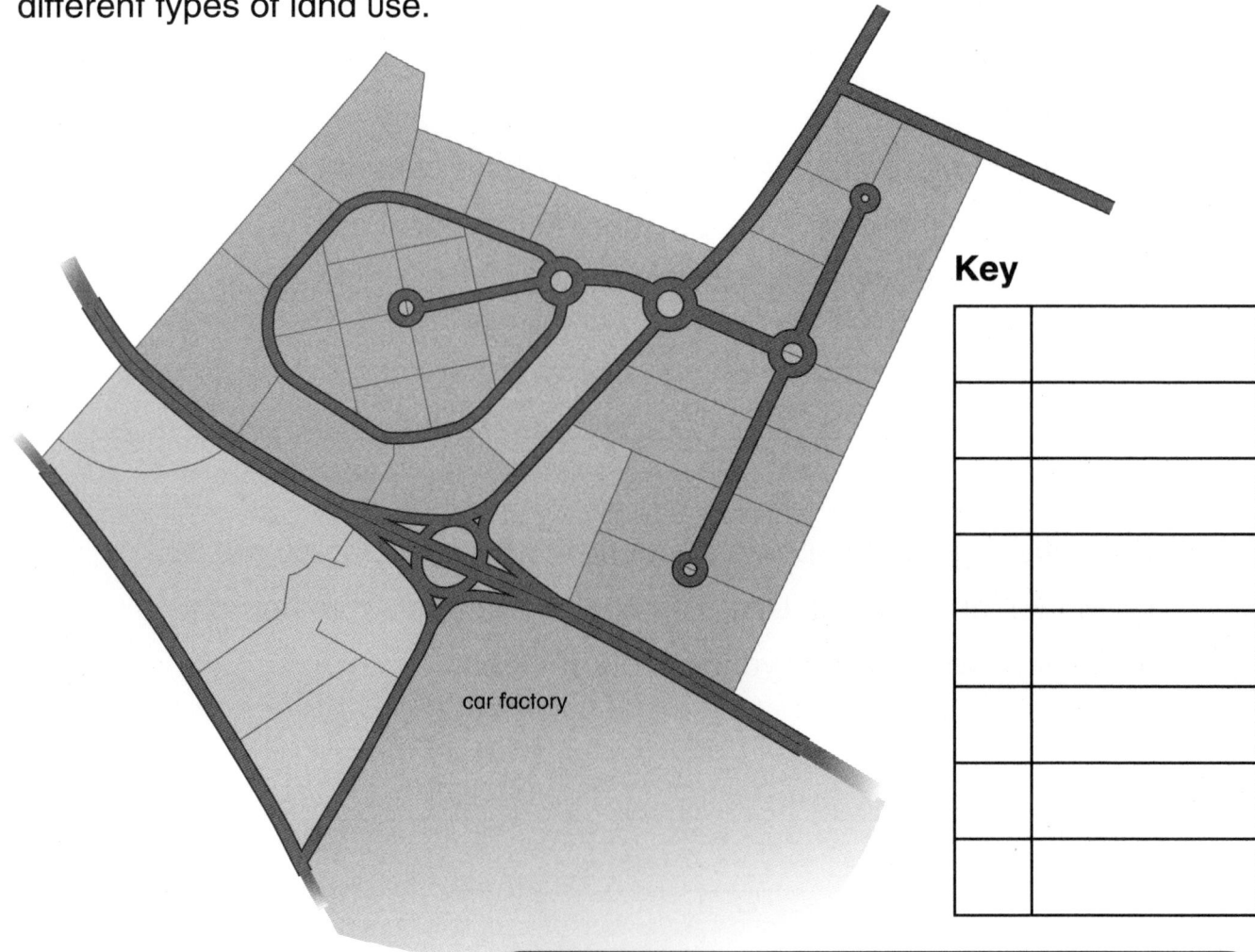

car factory

Key

→ Supports Pupil Book Investigation, page 23

Unit 4 Planning issues

3 Imagine that your school has moved to a new site. You are designing a new school which offers the same facilities but has a new, fresh design.

   a) First, make a rough map on a separate sheet of paper of where different facilities are now: classrooms, toilets, staffroom, hall, sports fields, and so on.

   b) Then think carefully about how you would improve the layout and the space in the rooms. Add some notes to your separate sheet of paper.

   c) Finally, draw your new school plan here. Label each area, and give your plan a title.

Unit 4 — Planning issues

Lesson 3: Planning game

1 a) Imagine you are a developer. You want to buy the site of an old school that is no longer being used. You want to redevelop it in a way that will be good for the local community.

What does your community need? Tick ✓ your choices.

- more housing ☐
- shops ☐
- medical facilities ☐
- sports and leisure ☐
- factory or farming ☐
- technology ☐
- small businesses ☐
- solar panels ☐
- park or woodland ☐
- other: _____ ☐

b) Choose which ones you want to develop. Remember, they need to fit into the size of an old school site. Explain your choices.

_____

_____

_____

c) Draw a rough plan of the site with the new development. Will you use the old school building or rebuild?

→ Supports Pupil Book Mapwork, page 25

Unit 4   Planning issues

2 Look at this aerial view of a residential area. What features are labelled? Write what they are and why they are good for the local community. Use the words from the boxes to help you.

(tennis court) (park) (train station) (main road) (swimming pool)
(house) (office block) (car park)

1 _____
2 _____
3 _____
4 _____
5 _____

25

Unit 5 Transport

Lesson 1: Travelling further, travelling faster

1 a) How many different types of transport can you think of? Write a list.

_____   _____   _____

_____   _____   _____

b) Which types of transport do you think are:

   i   fastest _____

   ii good for the environment _____

   iii safest _____

   iv best for local journeys in your town _____

   v   best for travelling across your country _____

2 Plan three different journeys from Rome, Italy, to Dublin, Ireland. Mark the stages of the journeys on the map.

   • Use a different colour for each stage (car, bus, train, ship, plane, and so on).

   • Make a key of the colours showing the different modes of transport.

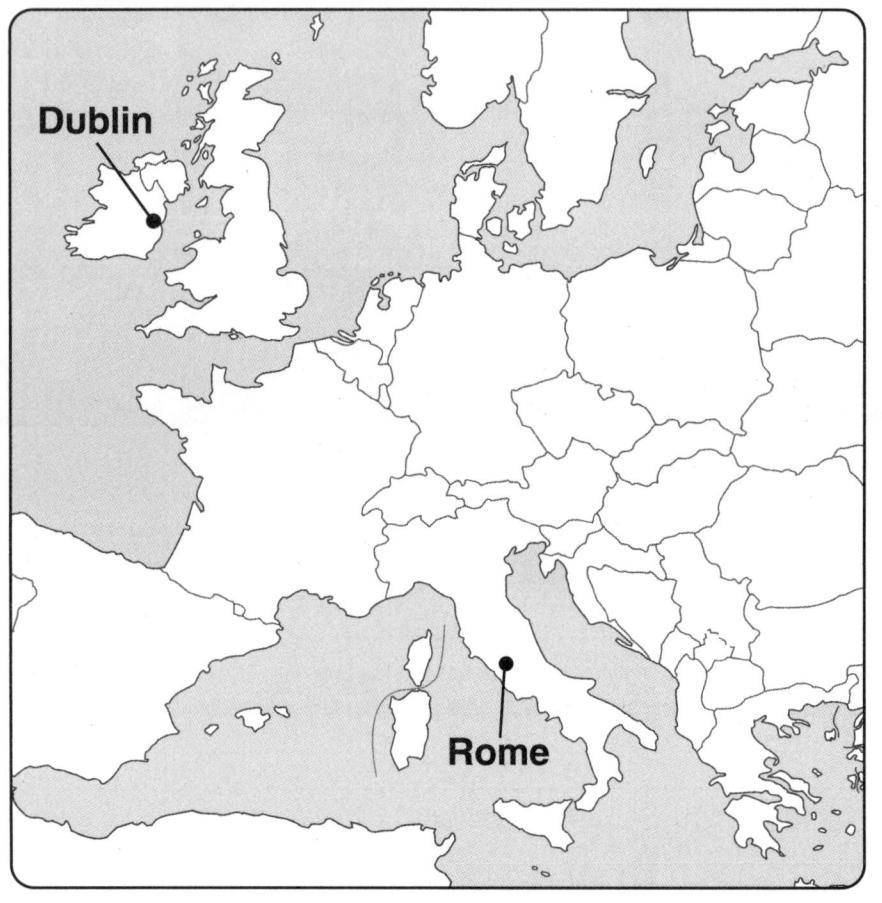

Key

Unit 5 Transport

3 a) Give the advantages and disadvantages of each form of transport in the table below.

b) Circle a score for each form of transport, in terms of cost, convenience and impact on the environment.

Form of transport	Advantages	Disadvantages	Score
car			Cost: 1 2 3 4 5 Convenience: 1 2 3 4 5 Impact on environment: 1 2 3 4 5
bus			Cost: 1 2 3 4 5 Convenience: 1 2 3 4 5 Impact on environment: 1 2 3 4 5
train			Cost: 1 2 3 4 5 Convenience: 1 2 3 4 5 Impact on environment: 1 2 3 4 5
boat			Cost: 1 2 3 4 5 Convenience: 1 2 3 4 5 Impact on environment: 1 2 3 4 5
plane			Cost: 1 2 3 4 5 Convenience: 1 2 3 4 5 Impact on environment: 1 2 3 4 5

Unit 5 Transport

Lesson 2: Transport problems

1 **a)** Add traffic lights, roundabouts, stop signs or one-way streets to help traffic run smoothly through this small section of road.

b) Show the direction of travel with arrows.

c) Plot a journey from home to the bus station and draw the route on the map.

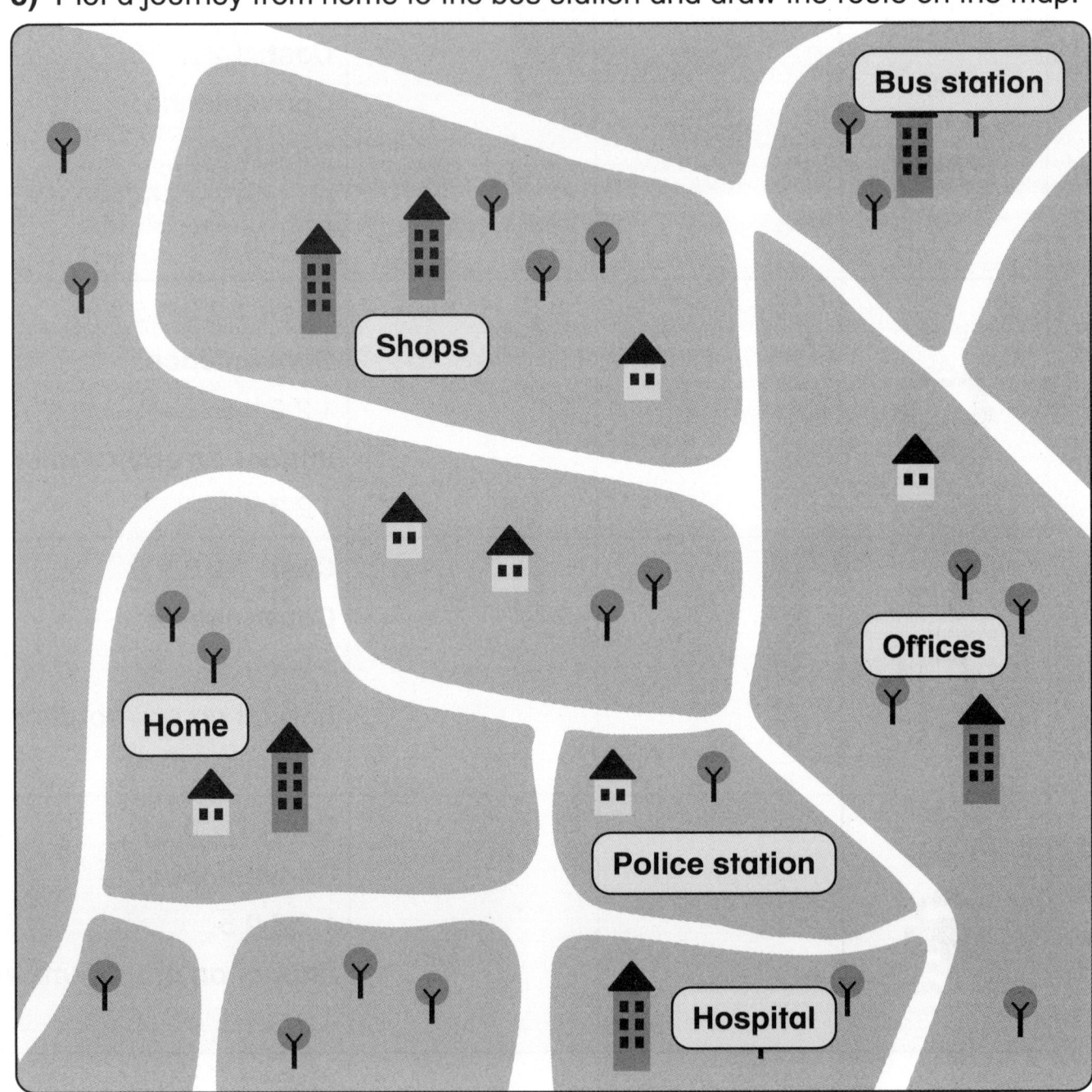

→ Supports Pupil Book Mapwork, page 29

Unit 5 Transport

2 Look at the pictures. What traffic problems do you think each one provides solutions for? Write your ideas below each picture.

Unit 5 Transport

Lesson 3: Hidden costs

❶ Which problems are the most serious in your area? Choose four.
Tick ✓ your choices.

> **Hint:** Use your survey results if you have completed the Investigation on page 31 of your Pupil Book.

Noise from traffic ☐

Not enough crossing places ☐

Traffic travelling too fast ☐

Roads with no pavements ☐

Not enough safety barriers ☐

Too many parked cars ☐

Shortage of cycle routes ☐

Exhaust fumes ☐

Heavy lorries ☐

Rush hour traffic jams ☐

→ Supports Pupil Book Investigation, page 31

❷ Suggest a solution for each of your four problems.

_____

_____

_____

_____

Unit 5 Transport

❸ You are going to design a vehicle that travels on or above land.

These are the requirements:
- It must seat 6 passengers comfortably.
- The passengers need to be safe in the vehicle.
- It needs to run on environmentally friendly fuel.
- It will run in a bus lane (or above it).
- It needs to help solve the problem of too much traffic.

Hint: Be inventive! Your vehicle can be realistic, or it can be futuristic, like a flying bus!

Give your vehicle a name, label the parts and add any notes you think will help the viewer to understand how your vehicle works.

Unit 6 Conservation

Lesson 1: Threatened wildlife

❶ a) Find out about a creature that is endangered in your country and complete the fact file.

b) Stick in a photo of the animal or draw a picture in the space.

Endangered! _____
What sort of creature is it?   _____
Where is the creature from, and what sort of habitat does it live in?   _____
Why is the creature endangered?   _____
What can be done to help?   _____

➜ Supports Pupil Book Investigation, page 33

Unit 6 Conservation

2 Test your conservation terms by finding these words in the wordsearch puzzle.

- endangered
- hunting
- species
- rhino
- extinct
- pesticides
- medicines
- orchid
- pollution
- threat
- mahogany
- whale

e	x	t	i	n	c	t	l	d	a	y
n	e	l	a	h	w	h	p	i	k	n
d	e	w	z	m	u	r	e	h	r	a
a	s	p	e	c	i	e	s	c	h	g
n	w	r	v	q	g	a	t	r	i	o
g	p	o	l	l	u	t	i	o	n	h
e	f	c	n	d	f	p	c	y	o	a
r	q	s	e	n	i	c	i	d	e	m
e	b	x	a	j	j	h	d	o	t	v
d	z	s	k	p	m	c	e	d	i	b
h	u	n	t	i	n	g	s	y	t	e

Hint: Some of the words are backwards!

3 Choose three of the words from the wordsearch puzzle. Write a sentence with each word to show that you understand what it means.

_____

_____

_____

_____

_____

Unit 6 Conservation

Lesson 2: Antarctica

1 Antarctica is a special continent covered by a huge sheet of ice. The ice in Antarctica has been melting over the years.

a) Read through all the information on pages 34 and 35 of your Pupil Book.

b) Fill in the boxes with information about why the ice should be protected, and what is threatening this area.

c) Use these words in your boxes:

- last wilderness
- research
- long ago
- pollution
- fresh water

Unit 6   **Conservation**

❷ Imagine you are Captain Scott. You are going to write a diary entry about your journey to the South Pole.

a) Use a book or the internet to find out about Captain Scott's journey to the South Pole.

    i When did Scott and his team make their journey?

    _____

    ii What animals did they bring with them?

    _____

    iii Where did they sleep and what did they wear?

    _____

    iv What animals did they study?

    _____

    v Did they make it to the South Pole? Were they the first?

    _____

b) Now write the diary entry describing what you did, how you travelled, what you saw, some of the difficulties you had and how you felt.

_____

_____

_____

_____

_____

_____

_____

Captain Scott

→ Supports Pupil Book Investigation, page 35

Unit 6 Conservation

Lesson 3: Conservation projects

1 Complete this information leaflet about the conservation of monarch butterflies.

   a) Colour in the butterfly to show the markings of a monarch butterfly.

   b) Write short notes under each heading.

About monarch butterflies

_____
_____
_____
_____

The route they take

_____
_____
_____
_____

Dangers to the butterflies

_____
_____
_____

Conservation/What happened

_____
_____
_____
_____
_____
_____
_____

Unit 6 Conservation

2 How could your school or local park be improved to help wildlife in the area?

> **Hint:** Wildlife refers to all creatures, including butterflies, snails and flowers, not just big creatures like tigers!

a) Make a rough plan of your school or your local park.

b) Draw three or more examples of improvements on your plan.

c) Describe your improvements in short notes.

_____
_____
_____
_____
_____
_____
_____

→ Supports Pupil Book Mapwork, page 37

Unit 7 England

Lesson 1: Introducing England

1 a) Mark the following features on the map of England:
- the cities from the map on page 39 of your Pupil Book
- the border between England and Scotland in blue
- the border between England and Wales in red.

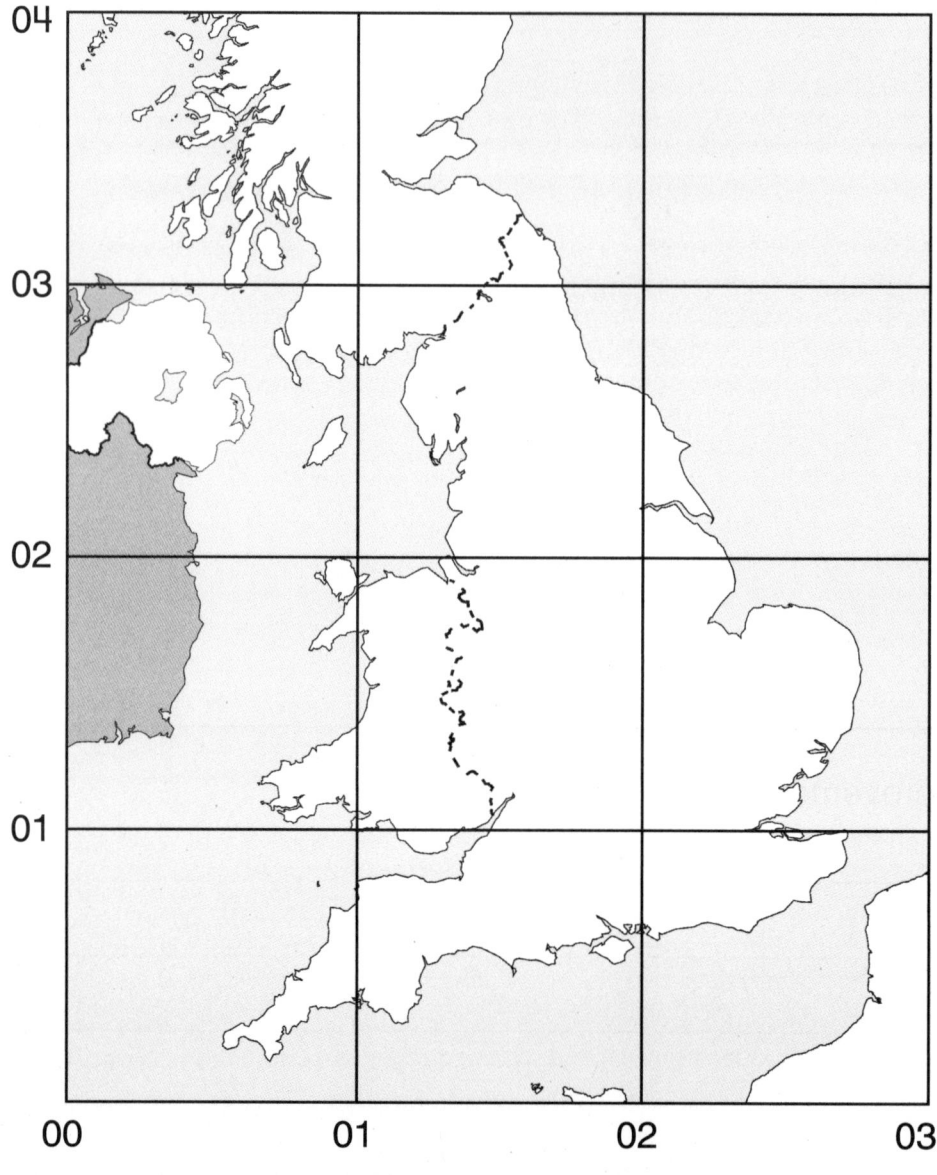

Hint: Remember to go across then up when reading grid references (along the corridor and up the stairs).

b) Identify the grid references for the cities.

London – Square 02,01 _____ _____

_____ _____ _____

_____ _____ _____

_____

Unit 7   England

2 Make a list of as many English cities as you can, arranging them in order from north to south.

_____
_____
_____
_____
_____
_____

3 a) Choose two topics from this list. Colour your choices.

- Weather
- Settlement
- Transport
- Work
- Rivers and landscape

b) Write three sentences about each topic and how it relates to England. Find or draw pictures to support your text.

Unit 7 England

Lesson 2: Finding out about Sandwich

❶ Reread pages 40 and 41 of your Pupil Book and look at the photographs.

❷ Make a timeline showing how Sandwich has developed and changed in the past 700 years.

- On the left of the timeline, write a description of what happened.
- On the right, draw small drawings or symbols to illustrate each event.
- Look ahead to page 42 of your Pupil Book. Add the new bypass and the Discovery Park to your timeline. Find out the dates or use approximate answers.

Hint: The 13th century is the years 1201–1300. The 14th century is the years 1301–1400. And so on.

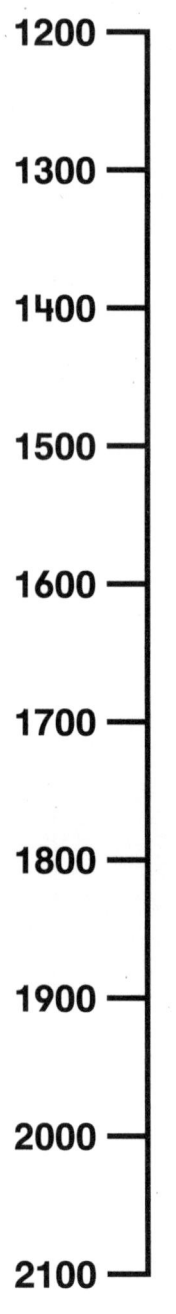

Unit 7 England

3 a) Look at the map of Sandwich from page 41 of your Pupil Book. Label the map with features from the boxes.

- old coal-fired power station
- science park
- railway
- bypass
- renewable energy plant
- River Stour

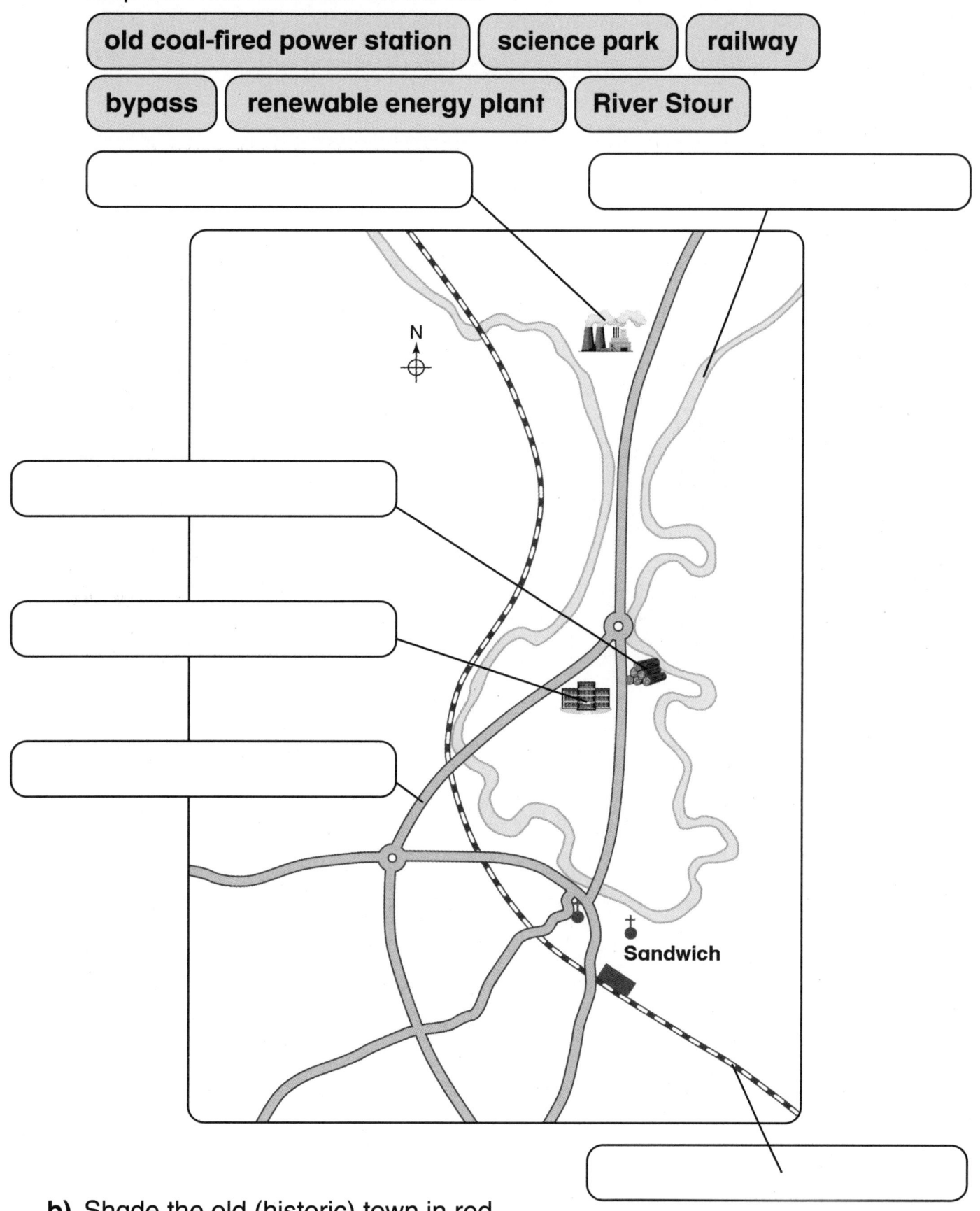

b) Shade the old (historic) town in red.

c) Shade the modern town in orange.

Unit 7 England

Lesson 3: Living in Sandwich

1 a) Make a list of four new developments in Sandwich.

b) Tick ✓ the options that apply to each new development.

New development	
	Protects the environment ☐ Creates jobs ☐ Protects the town centre ☐
	Protects the environment ☐ Creates jobs ☐ Protects the town centre ☐
	Protects the environment ☐ Creates jobs ☐ Protects the town centre ☐
	Protects the environment ☐ Creates jobs ☐ Protects the town centre ☐

2 Find out what people think about living in your area.

- Think of three questions to ask.
- Find at least six people of different ages and ask them the questions.
- Record their answers on a separate sheet of paper.
- In two clear sentences, summarise what people like and don't like about living in your area.

→ Supports Pupil Book Investigation, page 43

Unit 7 England

❸ Design a short walk around your local area for a child. Draw a map of the walk area in the box below.

   a) Mark five or six areas of interest on your map.

   b) Show the route with a dotted line and arrows.

   c) Mark your starting point, and the place where the walk ends.

   d) Illustrate your walk with photographs or draw pictures.

Unit 8 Europe

Lesson 1: Introducing Europe

What are the regions of Europe?

1 You are going to mark the mountains, grasslands, forests, Mediterranean landscape and tundra on this map of Europe.

a) Add symbols and colours to the key.

b) Now mark the different landscapes on the map, using the colours and symbols from the key.

mountain	grassland	forest	Mediterranean	tundra

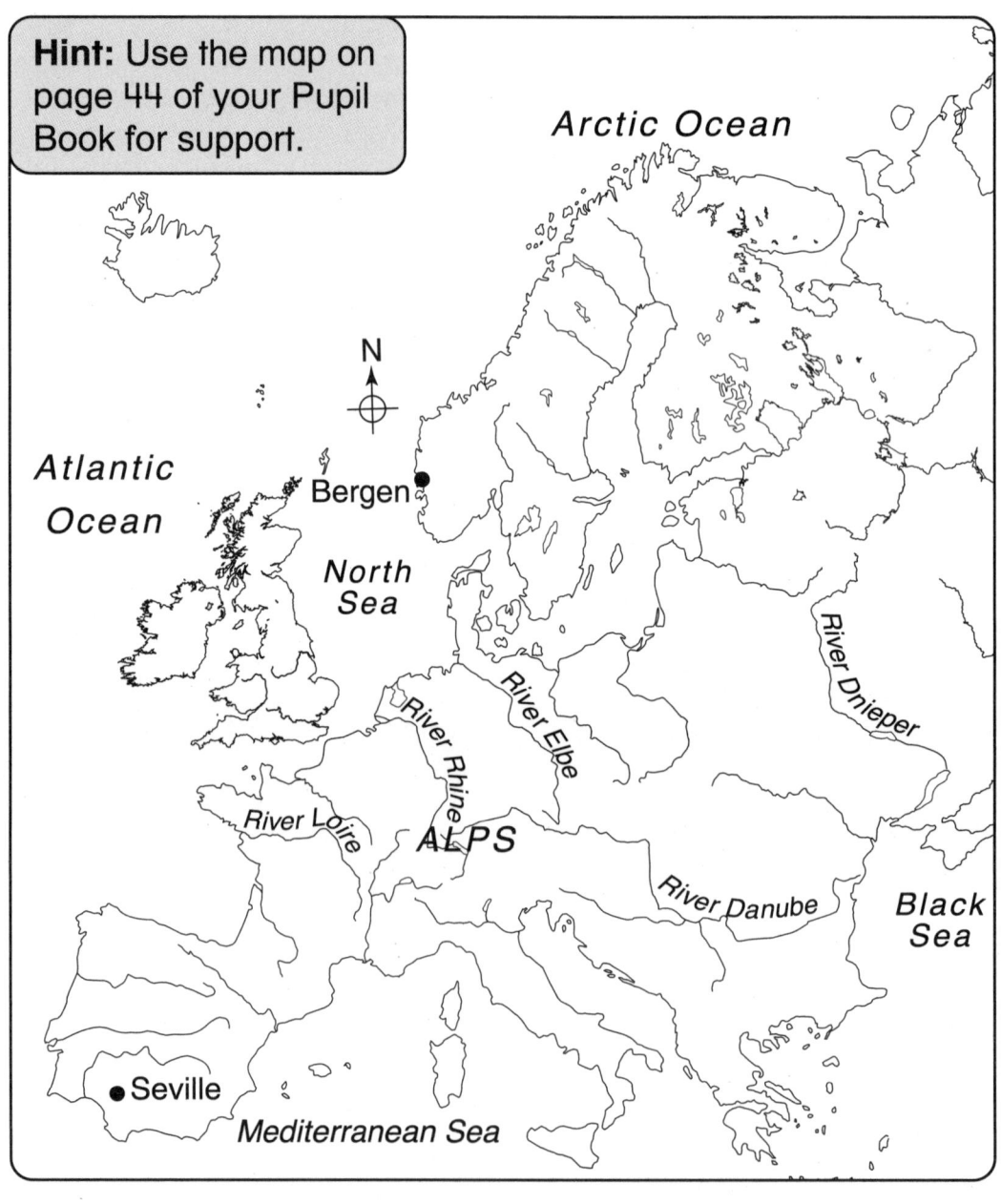

Hint: Use the map on page 44 of your Pupil Book for support.

44

Unit 8 Europe

2 Imagine you are travelling from Seville, Spain, to Bergen, Norway by car and ferry.

a) Use an atlas to find and name the countries you might travel through.

b) Pick four countries you might travel through. Write a sentence about each one.

c) Add a photograph or a drawing to illustrate something about each country.

→ Supports Pupil Book Investigation, page 45

Unit 8 Europe

Lesson 2: The European Union

1 a) List the countries which belonged to the EU in 1957 and 2025.

  b) Colour the keys and the maps to show the countries which belonged to the EU on each date.

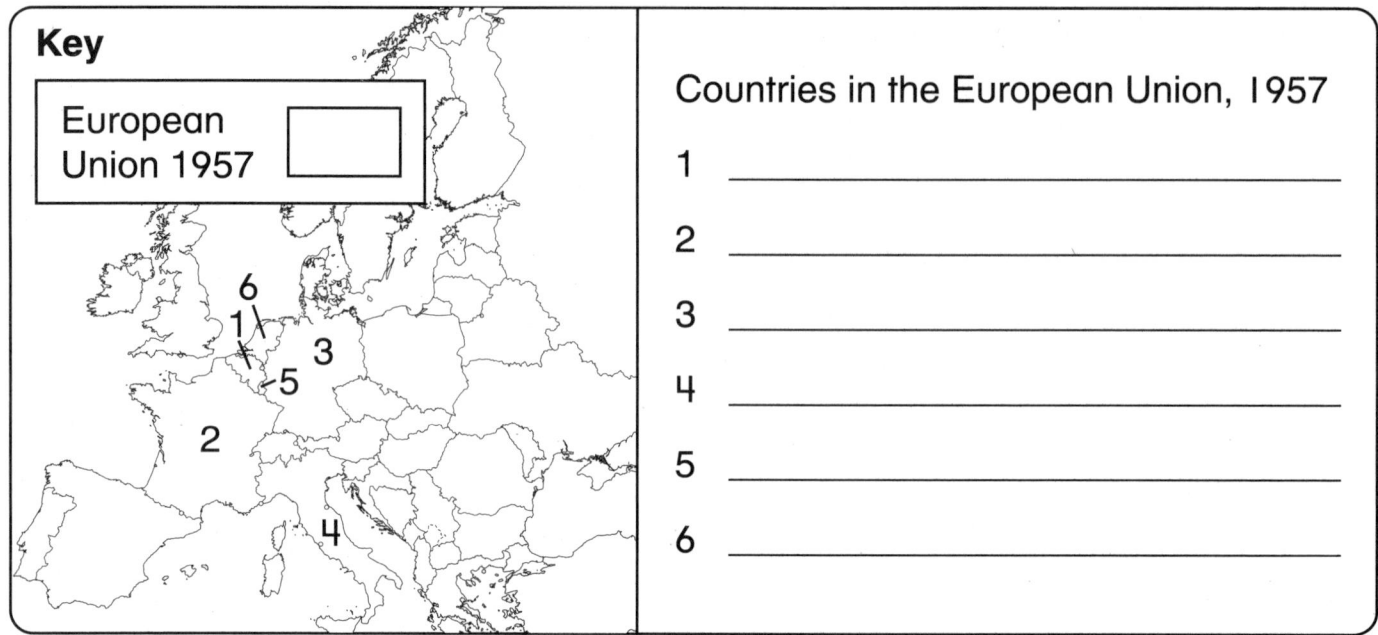

Key

European Union 1957 ☐

Countries in the European Union, 1957

1 _____
2 _____
3 _____
4 _____
5 _____
6 _____

Key

European Union 2025 ☐

Countries in the European Union, 2025

1 _____ 15 _____
2 _____ 16 _____
3 _____ 17 _____
4 _____ 18 _____
5 _____ 19 _____
6 _____ 20 _____
7 _____ 21 _____
8 _____ 22 _____
9 _____ 23 _____
10 _____ 24 _____
11 _____ 25 _____
12 _____ 26 _____
13 _____ 27 _____
14 _____

Unit 8 Europe

❷ Read the information on page 47 of your Pupil Book.

Complete this spider diagram, summarising all the ways in which the EU helps countries to work together.

Unit 8 Europe

Lesson 3: Celebrating Europe

❶ Choose three European countries.

    a) Write the name of the countries in the boxes in row 1.

    b) Draw and colour the flag of the three countries in the boxes in row 2.

    c) Draw and label a famous monument from each country in row 3.

    d) Write 'hello' in the language of each country in row 4.

    e) Write an interesting fact about each country in row 5.

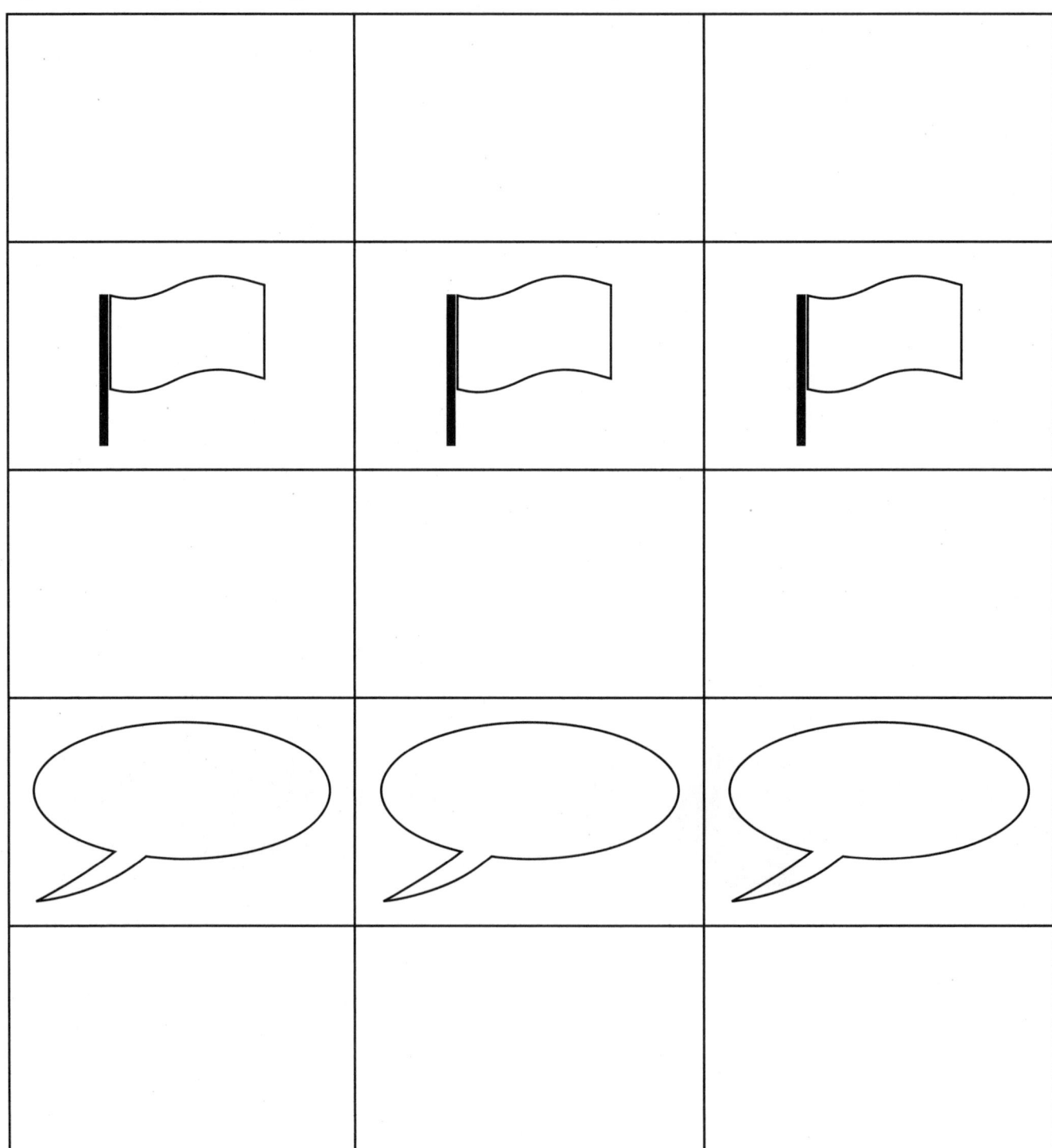

Unit 8   Europe

② Use pages 44 to 49 of your Pupil Book to write five interesting questions about Europe and the European Union.

a) Write your questions here, leaving space for the answers.

Question 1: _____

Answer: _____

Question 2: _____

Answer: _____

Question 3: _____

Answer: _____

Question 4: _____

Answer: _____

Question 5: _____

Answer: _____

b) Now swap your book with a friend and answer each other's questions.

Unit 9 South America

Lesson 1: Introducing the Amazon

1 Make a fact file about the Amazon.

   a) Draw or stick in a map and a picture from the Amazon in the top two spaces.

   b) Choose four headings for your fact file, including 'Location and size'.

   c) Complete your fact file information.

Fact file: The Amazon	
Location and size:	

→ Supports Pupil Book Investigation, page 51

Unit 9   South America

2 Look carefully at the map of the Amazon basin at the bottom of the page.

a) How many countries does the River Amazon and its tributaries pass through? ☐

Hint: You might not need all of these lines.

b) Make a list of the countries.

_____   _____

_____   _____

_____   _____

_____   _____

_____   _____

→ Supports Pupil Book Mapwork, page 51

c) Imagine you are exploring the River Amazon in a canoe. Colour the route you would take on the map. Where would you start?

Unit 9 — South America

Lesson 2: Using the rainforest

1 Look at this diagram of the layers of a natural rainforest. Use the words from the boxes to label each part of the rainforest.

- thick leaves in the **understory**
- shrubs growing on the **forest floor**
- **canopy** protects the plants and animals below
- roots
- rain clouds

2 What happens to the rainforest if the trees get cut down? Draw three ideas.

Unit 9 **South America**

❸ Write an article about how the rainforest is being used.

- The photograph showing deforestation will go with your article.
- Give your article a big, bold heading.
- Decide whether you are writing about good or bad use of the rainforest.
- Make rough notes on a separate sheet of paper before you write your article.
- Write at least three paragraphs.

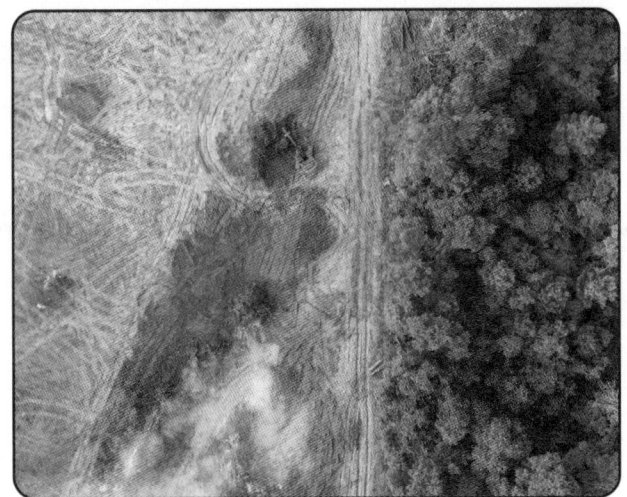

_____
_____
_____
_____
_____
_____
_____
_____
_____
_____
_____
_____
_____
_____
_____

Unit 9 | South America

Lesson 3: Saving the Amazon

1 Design a poster about why the Amazon rainforest needs protecting.

- Use some of the information on pages 52 to 55 of your Pupil Book about why the Amazon rainforest is important.
- Give your poster a big, bold title that people will notice.
- Draw pictures to complete your poster.

Unit 9 — South America

2 a) Test your Amazon vocabulary! Find these words in the wordsearch puzzle.

- basin
- tributary
- logger
- extinct
- farming
- tappers
- reserve
- soil
- species
- canopy
- forest

t	r	i	b	u	t	a	r	y	n
s	e	c	a	y	a	z	s	e	l
e	s	a	s	p	p	l	p	x	i
r	e	n	i	o	p	m	e	t	o
o	r	o	n	n	e	f	c	i	s
f	v	p	f	a	r	m	i	n	g
y	e	y	f	c	s	v	e	c	x
r	e	g	g	o	l	n	s	t	p

Hint: Some of the words are backwards!

b) Choose three of the words from part **a)** and use them in sentences of your own, to show that you know what they mean.

_____

_____

_____

Unit 10   Asia

Lesson 1: Southeast Asia

1 a) Colour the map of Southeast Asia, using a different colour for each country.

b) Label each country, using the names from the boxes.

Myanmar (Burma)   Thailand   Vietnam   Cambodia

Laos   East Timor (Timor-Leste)   Malaysia   Singapore

Brunei   Indonesia   Philippines

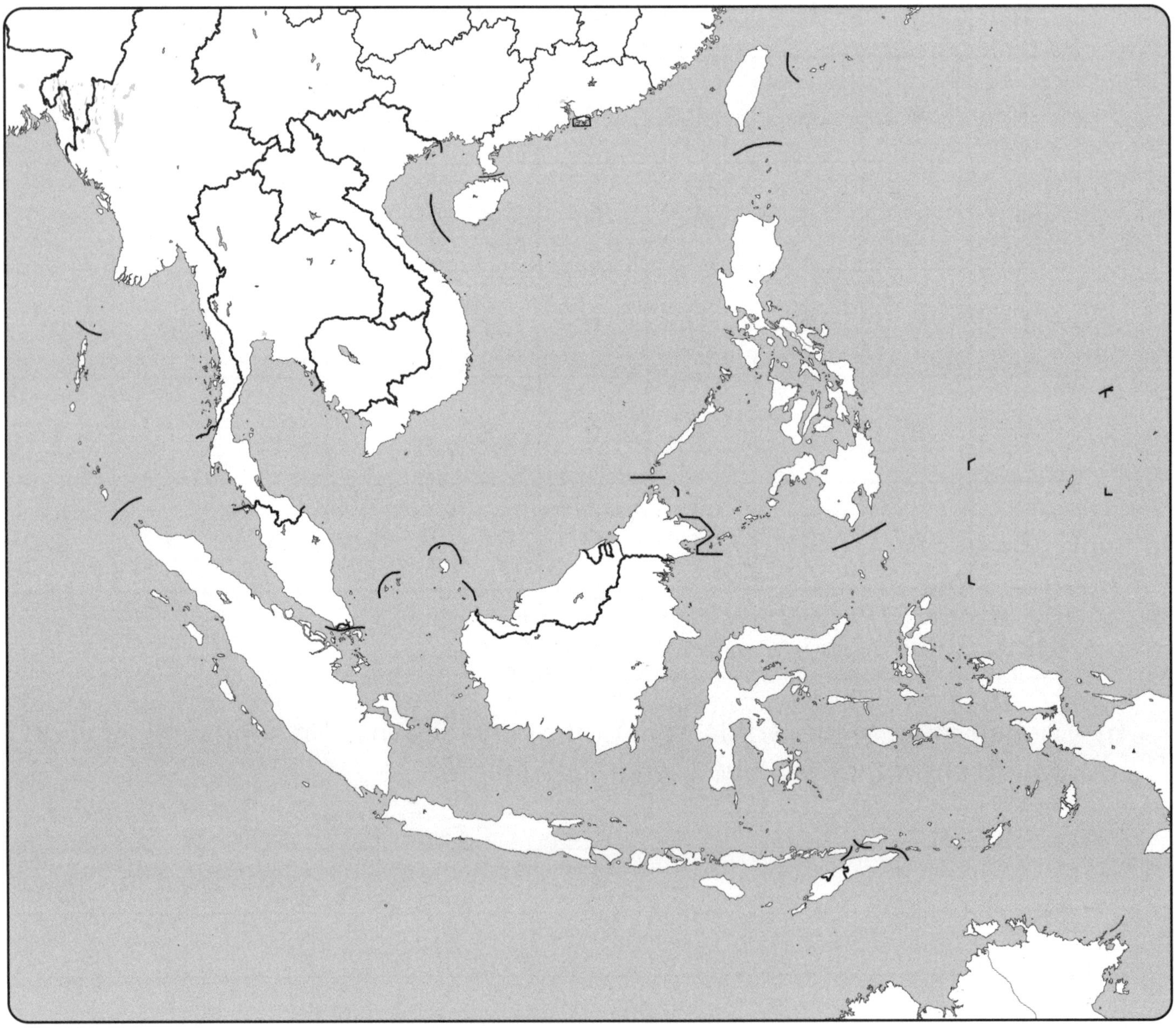

Hint: Refer to the map on Pupil Book page 56.

Unit 10 Asia

❷ Make a fact file for one Southeast Asian country, but do not choose Singapore. Add illustrations or photographs to the bottom boxes of your fact file and label them.

Country's name:

Landscape	Major cities

Products the country produces	Natural environment

→ Supports Pupil Book Investigation, page 57

Unit 10 | Asia

Lesson 2: Investigating Singapore

1 Write captions for these photographs. In each caption say:
- what the photo shows
- any other interesting information that is relevant.

_____
_____
_____

_____
_____
_____

_____
_____
_____

_____
_____
_____

Unit 10 Asia

❷ Label this outline map of Singapore, using the words from the boxes.

Hint: Look at the information on Pupil Book page 58.

bridge   docks   city centre   airport   Strait of Singapore

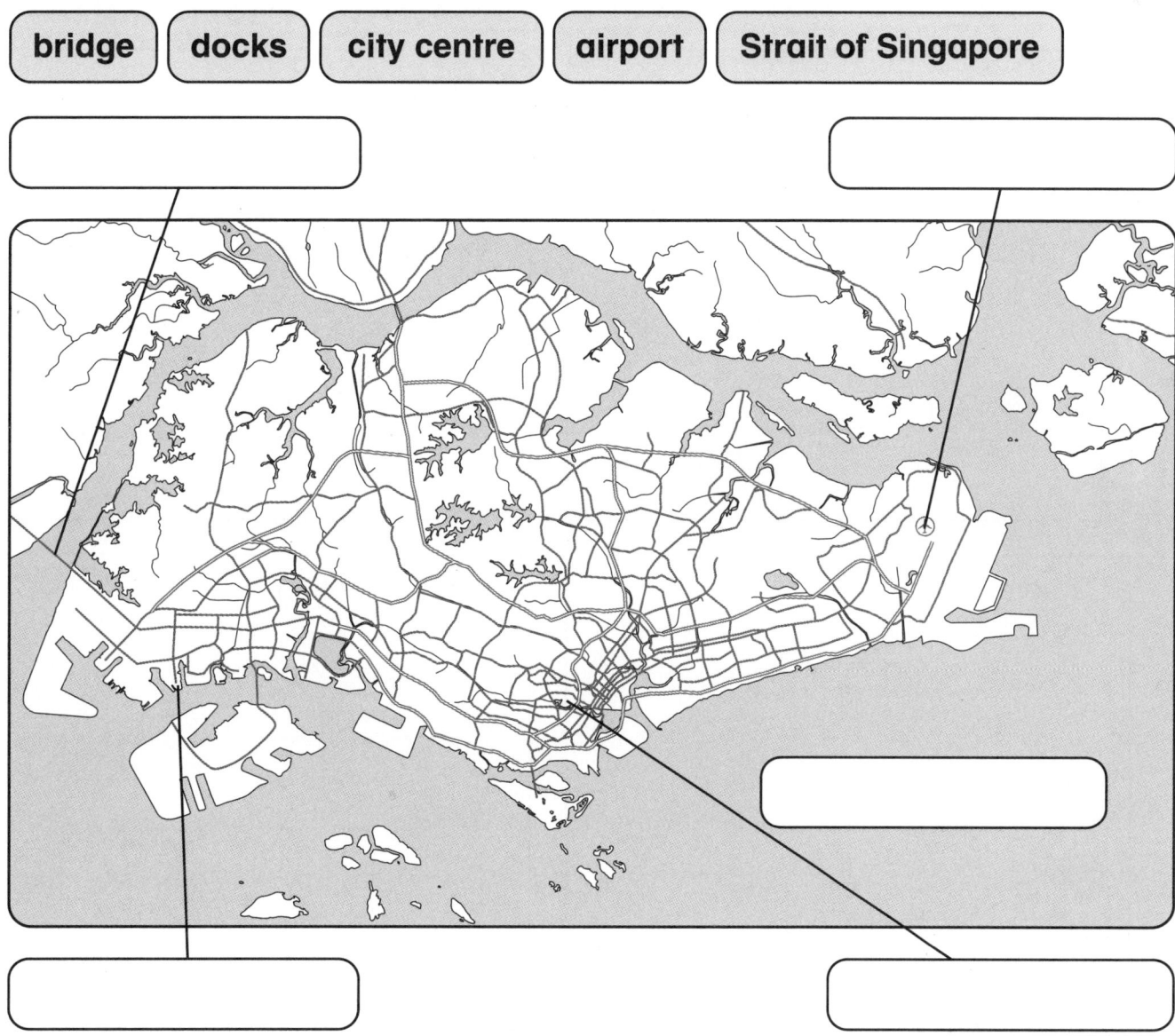

❸ Which two places would you like to visit in Singapore? Give reasons for your answers.

_____

_____

_____

_____

Unit 10 Asia

Lesson 3: A Singapore family

❶ Make a data and information sheet about Singapore using the headings in the boxes.

Map	History

Climate	Communication

Trade and industry	Environment

Unit 10   Asia

2 Read the information on page 61 of your Pupil Book.

- Draw a plan to show you could arrange six blocks of flats around a precinct.
- The shared area could have a garden, a park, a fountain, shops, a café and meeting places.
- Draw all these aspects of the shared area on your plan.
- Show where paths would go between the blocks of flats (apartments).

→ Supports Pupil Book Mapwork, page 61

Notes

William Collins' dream of knowledge for all began with the publication of his first book in 1819.

A self-educated mill worker, he not only enriched millions of lives, but also founded a flourishing publishing house. Today, staying true to this spirit, Collins books are packed with inspiration, innovation and practical expertise.
They place you at the centre of a world of possibility and give you exactly what you need to explore it.

Published by Collins
An imprint of HarperCollins*Publishers*
The News Building, 1 London Bridge Street, London, SE1 9GF, UK

HarperCollins*Publishers*
Macken House, 39/40 Mayor Street Upper, Dublin 1, D01 C9W8, Ireland

Browse the complete Collins catalogue at
collins.co.uk

© HarperCollins*Publishers* Limited 2025
Maps © Collins Bartholomew 2025

10 9 8 7 6 5 4 3 2 1

ISBN 978-0-00-872839-7

All rights reserved. No part of this publication may be reproduced, stored in a retrieval system, or transmitted in any form by any means, electronic, mechanical, photocopying, recording or otherwise, without the prior written permission of the Publisher or a licence permitting restricted copying in the United Kingdom issued by the Copyright Licensing Agency Ltd, 5th Floor, Shackleton House, 4 Battle Bridge Lane, London SE1 2HX.

Without limiting the author's and publisher's exclusive rights, any unauthorised use of this publication to train generative artificial intelligence (AI) technologies is expressly prohibited. HarperCollins also exercise their rights under Article 4(3) of the Digital Single Market Directive 2019/790 and expressly reserve this publication from the text and data mining exception.

British Library Cataloguing-in-Publication Data

A catalogue record for this publication is available from the British Library.

Author: Fiona Macgregor
Publisher: Laura White
Product managers: Natasha Paul and Shelley Teasdale
Development editor: Judith Walters
Copyeditor: Catherine Dakin
Proofreader: Charlotte Christensen
Cover designer and illustrator: Steve Evans
Internal illustrator: Jouve India Private Ltd
Typesetter: David Jimenez
Production controller: Katie Jean-Baptiste
Printed and bound in the UK by Martins the Printers

This book is produced from independently certified FSC™ paper to ensure responsible forest management.

For more information visit: www.harpercollins.co.uk/green
collins.co.uk/sustainability

Acknowledgements

The publishers gratefully acknowledge the permission granted to reproduce the copyright material in this book. Every effort has been made to trace copyright holders and to obtain their permission for the use of copyright material. The publishers will gladly receive any information enabling them to rectify any error or omission at the first opportunity.

All photos: Shutterstock.